THE INTERNET

NANCY DICKMANN

BROWN BEAR BOOKS

Published by Brown Bear Books Ltd
4877 N. Circulo Bujia, Tucson, AZ 85718
USA
and
Studio G14, Regent Studios, 1 Thane Villas,
London N7 7PH, UK

Text: Nancy Dickmann
Design and Illustrations: Square and Circus
Design Manager: Keith Davis
Children's Publisher: Anne O'Daly

Library of Congress Cataloging-in-Publication Data
Names: Dickmann, Nancy, author.
Title: The internet / Nancy Dickmann.
Description: Tuscon, AZ, USA : Brown Bear Books Ltd, [2025] | Series: Science starters | Includes bibliographical references and index. | Audience: Ages 6-9 | Audience: Grades 2-3 | Summary: "Find out what the Internet is, how it works, and how to stay safe when using it"– Provided by publisher.
Identifiers: LCCN 2023053004 (print) | LCCN 2023053005 (ebook) | ISBN 9781781219904 (library binding) | ISBN 9781781219966 (paperback) | ISBN 9781835720028 (ebook)
Subjects: LCSH: Internet–Juvenile literature.
Classification: LCC TK5105.875.I57 D525 2025 (print) | LCC TK5105.875.I57 (ebook) | DDC 004.67/8–dc23/eng/20231122
LC record available at https://lccn.loc.gov/2023053004
LC ebook record available at https://lccn.loc.gov/2023053005

Picture Credits
The photographs in this book are used by permission and through the courtesy of:
Shutterstock: Brostock 10–11, Ronnakrit Kiratiarporn 16–17, Martin Lueke 8–9, Orion Production 14–15, PintoArt 18–19, sitthiphong 20–21, Tommy Lee Walker 12–13, Drazen Zigic 4–5, 6–7.
t-top, r-right, l-left, c-center, b-bottom

Words in **bold** appear in the Glossary on page 23.

Manufactured in the United States of America
CPSIA compliance information: Batch#AG/5660

CONTENTS

AN ONLINE WORLD

When's the last time you used the internet? Chances are, it wasn't more than a day or two ago. The internet is still pretty new. It only opened up to public users in 1993. But it has quickly become part of our everyday lives.

CONNECTING THE WORLD

The internet has made our world more connected. You can use it to learn about faraway places. You can chat with people in other countries. Want to watch videos or get instant news updates? Just use the internet! Without it, the world would be very different.

We connect computers to the internet. You can also connect using a phone.

One Web, Many Uses

The internet is really useful! We use it for many different things. These are just a few of them.

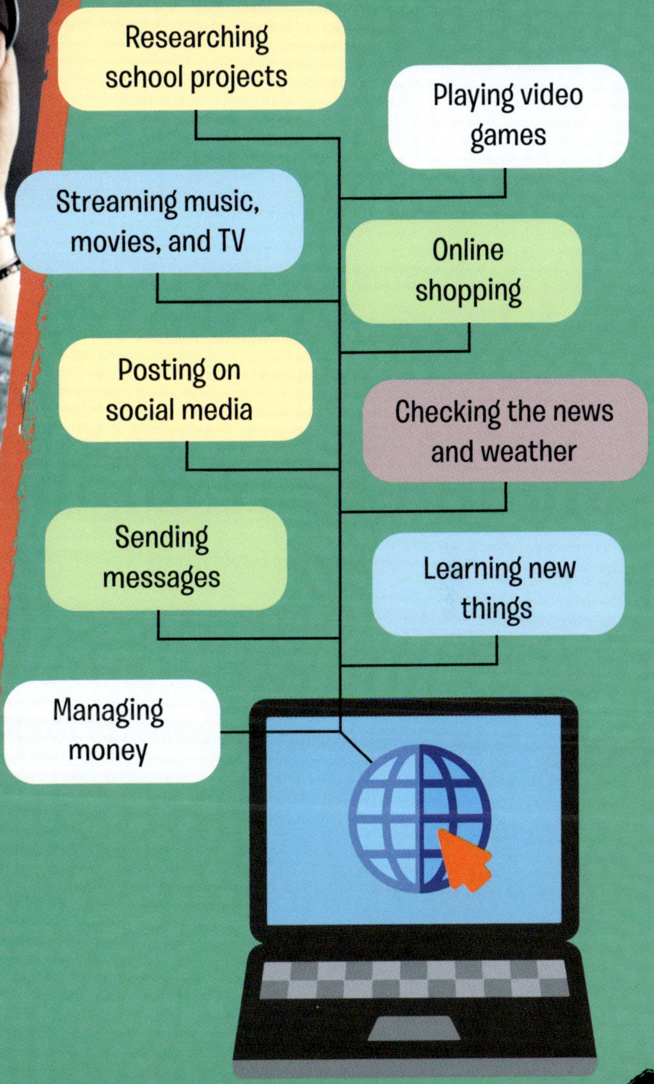

COMPUTER NETWORKS

One computer on its own is a useful tool. But it's even more useful if you can link it to another computer. Then they can "talk" to each other. They can share **files**. A group of linked computers is called a **network**. You might have one at home or at school.

No Wires Needed?

Sometimes the computers in a network are connected by **cables**. Data flows along the wires. Some **devices** can connect wirelessly. They send and receive signals through the air.

SUPER-SIZED

The internet is just a really, really big computer network! It links computers all over the world. They are able to share information. They send and receive **data**. Any computer can join. It just needs an **internet service provider** (ISP) to let it in.

The computers in a school network can share files. They can run the same **programs.**

Small Networks

Do you have a network at home? It might link different devices. They can all connect to the internet.

SENDING SIGNALS

Cable-laying ships are designed to carry long cables.

Most computers use wires to link up. The wires cover long distances. On land, they are often buried underground. Sometimes they hang from poles. But what if they need to cross an ocean? Special ships carry thick cables. They lay them on the sea floor.

UP IN THE AIR

Satellites can also send signals. Satellites fly high above the Earth. They send out invisible waves. A dish on the ground collects the signal. Not everyone has a cable near their home. They can use this system instead.

Mobile Internet

You can use the internet on a cell phone. You don't need wires. There are cell towers all over. The phone picks up a signal from the nearest one. It uses invisible waves.

Underwater Cables

The glass fibers that carry signals are easily damaged. They are surrounded by protective layers.

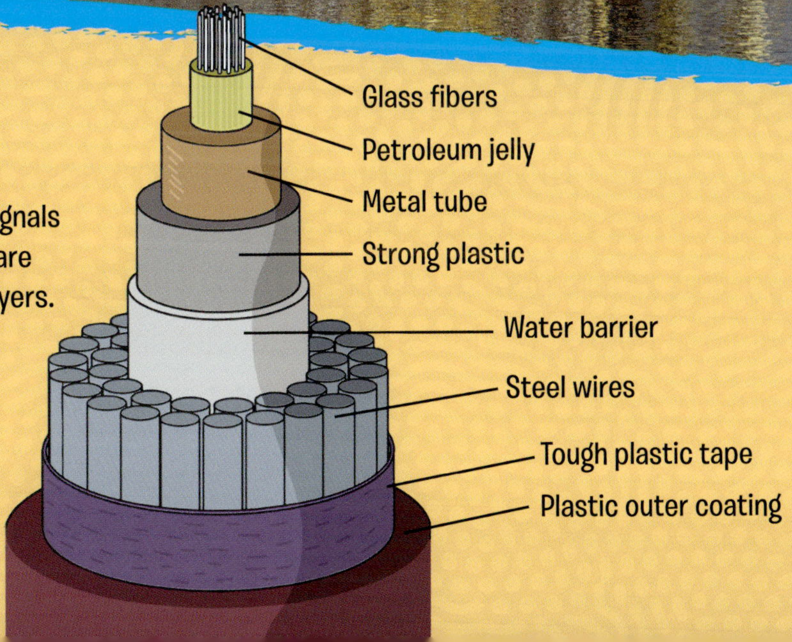

WEBSITES

Many internet users visit **websites**. Which one is your favorite? A website is a group of pages. They are linked together. The main page is called the home page. You can click links. They will take you to other pages.

WHAT'S IT FOR?

There are many different types of websites. Some report the news. Others are online shops. You can browse or buy. Some websites are forums where people can chat. Others are for sharing photos or videos. Many businesses have a website. It tells people what they do.

Websites let you check a restaurant's menu. You can read reviews and reserve a table.

Parts of a Website

A good website is easy to use. You can find your way around. Many websites have these features.

The URL is the website's address. You type it to go to the site.

Menus at the top help you go to another page.

A search box helps you find what you need.

The header has the website's name. It's always at the top.

The footer shows details about who made the site.

A sidebar has links to other websites.

HOW IT WORKS

You can click on a link to visit a website. You can type in the address. The website appears on your screen. Its text and images are stored on another computer. That computer might be far away. How does it all get to your screen?

SERVERS

"The internet" isn't a place. Everything on it is stored somewhere. That includes text, photos, and videos. All this data is stored on big, powerful computers. They're called **servers**. There are servers all over the world.

Data Centers

Servers are kept in a special building. It's called a data center. Servers get hot when they run. They need fans and air conditioning to stay cool. This uses a lot of electricity.

One server can store many websites. It stays on, day and night.

Making a Connection

To use the internet, you need a **browser**. It's a program that runs on your computer.

1. You click a link. You could also type an address.
2. Your ISP finds out where the website is stored.
3. It sends a request to the website's server.
4. The server sends a signal to your computer.
5. Your browser receives the signal. It shows the website on your screen.
6. You click a menu. This request goes to the server. It sends you the next page.

EMAIL

It's your friend's birthday! You write a message on a card. You need an envelope and stamp. It might take days to arrive. So why not send an email instead? It travels across the internet. It arrives in seconds!

EMAIL ADDRESSES

Do you have an email address? These addresses have two parts. The first is the user name. The second is the domain name. An "@" separates the two parts. Lots of addresses have the same domain. It's like a street with lots of houses. The user names are the individual houses. Each one is unique for that domain.

Some emails are just text. But you can also attach photos or files.

Parts of an Email

People send emails for work and school. They send them to friends and family. Every message has the same basic parts.

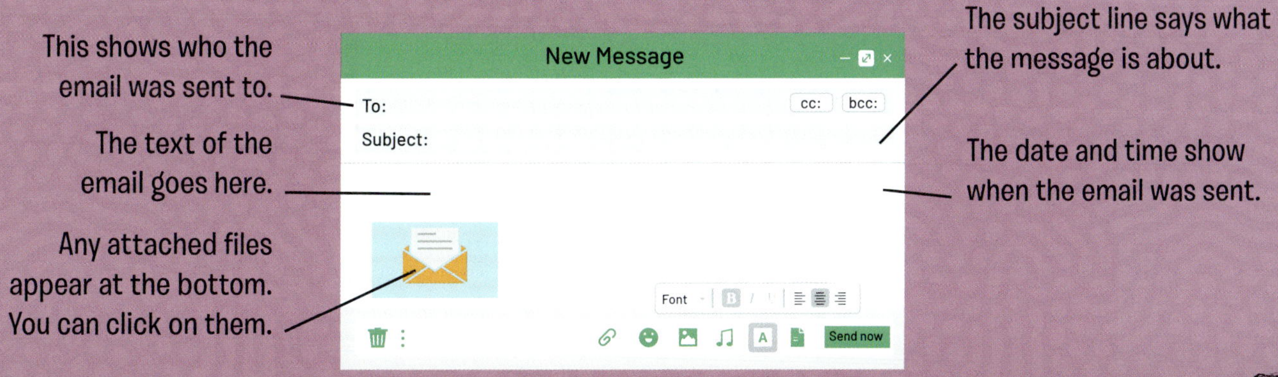

THE CLOUD

You've taken a great photo. You want to put it in a presentation. But the photo is on your phone. The presentation is on your laptop. How can you get the photo where you want? You could email it. Or you could upload it to the **cloud** instead.

SHARING FILES

The cloud is a network. It's made up of many servers. They are all over the world. You can save files on them. Then you can use them from any device. You connect to the servers through the internet. You might need to use a password. This keeps your files safe.

Real clouds float up in the sky. But that is not where your files are stored.

Using the Cloud

The cloud makes it easy to switch between devices. It lets people work together on the same document. Here's how it works.

You start writing a report on your tablet. It saves to the cloud.

The file is stored on a server somewhere.

Later, you're on a school computer. But you need to finish that report!

The school computer is connected to the internet. It can find your file.

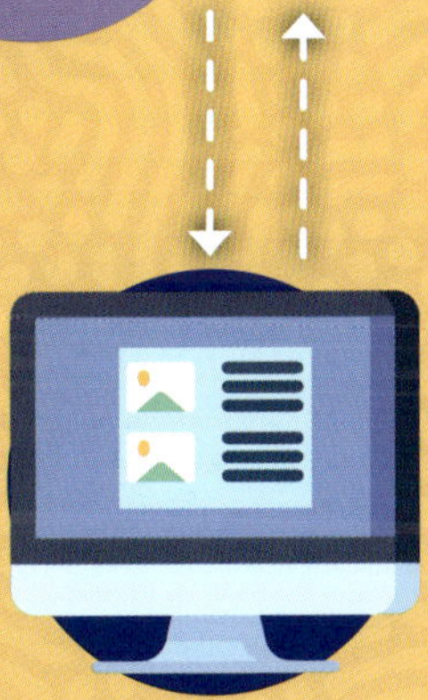

You save the finished version to the cloud again.

OUT AND ABOUT

Need help finding your way? A map app can help. It needs an internet connection to work.

Do you have an internet connection at home? It probably uses wires. They connect your home to the internet. Your devices might connect wirelessly. But the signal doesn't travel far. You have to be at home to use the internet.

MOBILE INTERNET

But what if you're not at home? You might need to look up an address. You might want to check the weather forecast. You can do this using mobile internet. It uses the same signals as cell phones do. It connects using invisible waves.

Paying for Data

You pay a cell company to use this service. You pay for the data you use. Sending an email uses very little. Streaming videos uses a lot more.

Passing It On

Cellular data uses the internet. So does your home computer. It's the same internet. You just connect in a different way.

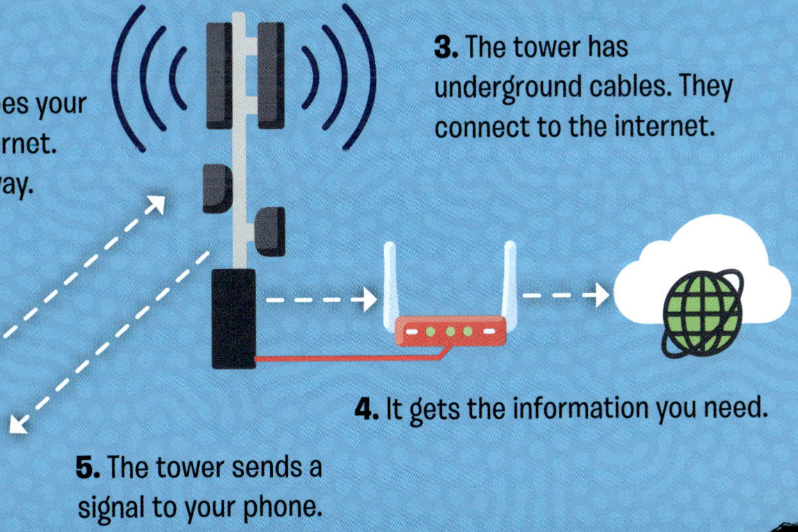

1. You need to be near a cell tower.

2. Your phone sends a request. It goes to the tower.

3. The tower has underground cables. They connect to the internet.

4. It gets the information you need.

5. The tower sends a signal to your phone.

STAYING SAFE ONLINE

The internet connects people all over the world. But some of those people are up to no good. You need to stay safe when using the internet. Make sure a grownup knows what websites you're using. Tell them if someone you met online wants to meet up.

When you're online, don't use your real name. Don't share your address.

KEEPING FILES SAFE

Hackers try to break into computers. They want to steal personal details. Delete emails if you don't know who sent them. Don't click on links or open files. They might damage your computer. An adult can help you stay safe online.

What Can I Do?

Here are some useful tips for staying safe online.

Treat others with respect. If you see bullying, tell an adult.

Don't give out personal details like your name or address.

Check your privacy settings. An adult can help.

Use strong passwords to protect files and accounts.

Think before you post. Who will be able to see it?

Don't believe everything people tell you.

QUIZ

How much have you learned about the internet? It's time to test your knowledge!

1. What is the name for a group of computers linked together?

a. association

b. club

c. network

2. How do internet signals cross the oceans?

a. through cables on the sea floor

b. across bridges

c. on submarines with special antennae

3. What is the name for a big, powerful computer that stores websites and other data?

a. supercomputer

b. server

c. control hub

4. What do you call the second part of an email address?

a. last name

b. user name

c. domain name

The answers are on page 24.

GLOSSARY

browser a computer program used for accessing and interacting with websites

cable a wire or bundle of wires with a protective casing, used for carrying electrical signals

cloud a worldwide network of servers that stores people's documents and programs

data information in the form of words or numbers

device a piece of electrical equipment such as a smartphone, tablet, or laptop

file a document made by a computer that contains information in the form of text, sound, or images

hacker a person who tries to break into computer systems

internet service provider (ISP) a company that provides its customers with access to the internet

network a group of computers connected together

program a set of coded instructions for a computer to follow

satellite an artificial object sent into orbit around Earth, which can send and receive communication signals

server a computer that stores and sends out information to other computers

website a group of linked web pages

wireless a way of connecting to a network using radio waves instead of wires

FIND OUT MORE

Books

Networks and the Internet. Nancy Dickmann, Gareth Stevens, 2019.

Staying Safe with Technology. Ashley Richardson, Capstone Press, 2022.

What Are Computer Networks and the Internet? Jennifer Reed, Rosen Publishing, 2018.

Websites

kids.britannica.com/kids/article/Internet/353293

www.bbc.co.uk/bitesize/topics/zs7s4wx/articles/z78nydm

www.dkfindout.com/us/computer-coding/what-is-world-wide-web/

INDEX

Answers: 1. c; 2. a; 3. b; 4. c